A

CBO

PAPER

Options for Combining the Navy's and the Coast Guard's Small Combatant Programs

July 2009

The Congress of the United States ■ Congressional Budget Office

Notes

Preface

As part of their long-term procurement strategies, the Navy and the Coast Guard are each in the process of developing and building two types of small combatants. The Navy is building two versions of its new littoral combat ship, and the Coast Guard is building replacements for its existing classes of high-endurance cutters and medium-endurance cutters. Although all four types of ship are about the same size, they are designed to perform different missions. If the Navy's and Coast Guard's plans for their small combatant programs are fully implemented, the two services combined will spend over $47 billion over the next 20 years purchasing 83 of those ships.

In light of the many pressures on the budgets of the Navy and the Coast Guard, some policymakers and analysts have questioned whether the services could combine their small combatant programs in ways that still meet their requirements but save money. This Congressional Budget Office (CBO) paper, prepared at the request of the Ranking Member of the Senate Budget Committee, examines three alternatives that might allow the Navy and the Coast Guard to consolidate their small combatant programs. In keeping with CBO's mandate to provide impartial analysis, the paper makes no recommendations.

Eric J. Labs of CBO's National Security Division wrote the paper under the general supervision of J. Michael Gilmore and Matthew Goldberg. Raymond Hall of CBO's Budget Analysis Division and Derek Trunkey of CBO's National Security Division prepared the cost estimates under the general supervision of Sarah Jennings and Matthew Goldberg. Paul Cullinan, Sarah Jennings, and David Moore provided thoughtful comments on a preliminary draft. Heidi Golding, Dawn Sauter Regan, and Jason Wheelock of CBO contributed to developing the life-cycle cost estimates used in the paper. Robert Work, formerly of the Center for Strategic and Budgetary Assessments, reviewed the manuscript before publication. (The assistance of external reviewers implies no responsibility for the final product, which rests solely with CBO.)

Loretta Lettner edited the paper, and Christine Bogusz proofread it. Maureen Costantino designed the cover and prepared the document for publication. Lenny Skutnik printed the initial copies, Linda Schimmel handled the print distribution, and Simone Thomas prepared the electronic version for CBO's Web site (www.cbo.gov).

Douglas W. Elmendorf
Director

July 2009

Contents

Tables

Figures

Boxes

Options for Combining the Navy's and the Coast Guard's Small Combatant Programs

Summary and Introduction

As articulated in their respective long-term shipbuilding plans, the Navy and the Coast Guard intend to spend more than $47 billion combined over the next 20 years to purchase a total of 83 small combatants. Of that number, the Navy plans to purchase 53 littoral combat ships (LCSs), in addition to the two that were purchased in 2005 and 2006. The LCSs will be built using two different hull designs—one, a semiplaning monohull; the other, an aluminum trimaran—although the exact mix of hulls has not yet been determined.[1] The ships will carry one of three sets of equipment, or mission packages, depending on which mission they are expected to perform (antiship, antisubmarine, or countermine warfare).

The Coast Guard plans to buy five new high-endurance cutters, commonly referred to as national security cutters (NSCs), and 25 new medium-endurance cutters, often called offshore patrol cutters (OPCs).[2] Three other NSCs ordered prior to 2009 have been built or are currently under construction. Although the Coast Guard plans to begin buying the offshore patrol cutter in 2015, it is not yet certain what the OPC will look like or if it will be

confined to one class of ship. Together, the NSCs and OPCs, which are designed to operate 50 nautical miles beyond the U.S. coastline, are part of the resources and force structure that make up the Coast Guard's "Deepwater assets."[3]

As the designation "small combatant" implies, the Navy's LCSs and the Coast Guard's NSCs and OPCs are designed to be significantly shorter in length, lighter in weight, and shallower in draft than most Navy surface warships (carriers, amphibious ships, cruisers, and destroyers). For instance, the Navy's LCSs have a full-load displacement—the weight of the ship plus its crew, weapons, fuel, and cargo—of about 3,300 tons. The national security cutter displaces about 4,300 tons of seawater, and the most recent notional design of the offshore patrol cutter would have the ship displace 3,700 tons.[4] By contrast, the most modern Arleigh Burke class destroyers, the backbone of the Navy's surface combatant force, displace about 9,500 tons; amphibious ships displace from 16,000 to 45,000 tons; and aircraft carriers displace about 100,000 tons.

Despite a superficial similarity in size, the small combatants being developed by the Navy and the Coast Guard

---

1. A team led by Lockheed Martin is building a steel, semiplaning monohull. A team led by General Dynamics is building an aluminum trimaran.

2. The Coast Guard's official designations for the ships are Maritime Security Cutter, Large (WMSL-750) and Maritime Security Cutter, Medium (WMSM). In this analysis, the term "national security cutter" will often be used to describe the Coast Guard's new high-endurance cutter and the term "offshore patrol cutter" will often be used to describe the Coast Guard's new medium-endurance cutter. The term "cutter" describes any Coast Guard vessel that is 65 feet or greater in length and can accommodate a crew for extended periods of time.

3. Other Deepwater assets include aircraft and sensors. The Coast Guard's Deepwater program is a 30-year plan to recapitalize virtually all of the service's ships, aircraft, and sensors that perform missions more than 50 nautical miles from U.S. shores.

4. All displacements are measured in long tons, which is typical for most Navy and Coast Guard ships. Displacements for the LCS, however, are usually expressed by the Navy in metric tons. To convert metric to long tons, multiply metric tons by 0.984. A standard short ton is 2,000 pounds; a long ton is 2,240 pounds; and a metric ton is 2,204.62 pounds.

have different characteristics and capabilities, which are designed to fulfill different missions. In general, the Coast Guard ships are meant to operate independently at sea for long periods of time and at some distance from the shore (that is, to have a large, unrefueled range of operation) and not to engage in major combat operations. The Navy's LCSs, by contrast, are designed to have less range than Coast Guard cutters but to operate at much greater speeds and serve during wartime as part of a naval battle network in close-in littoral waters in other parts of the world. Those differences in mission requirements have led the two services to reject a common ship platform for their small combatant needs and to adhere to their previously articulated procurement plans.

In the early stages of implementing those plans, however, the Navy and the Coast Guard have encountered various challenges. Cost overruns and construction problems have plagued both versions of the Navy's littoral combat ship—designated LCS-1 (the semiplaning monohull) and LCS-2 (the aluminum trimaran)—as well as the Coast Guard's national security cutter. Because of the difficulties associated with constructing the NSC, development of the Coast Guard's notional offshore patrol cutter has been delayed by five years. All three ship programs have experienced substantial cost growth above the original estimates the services provided to the Congress as well as delays in construction of more than a year for each ship.

As a result of those delays and cost overruns, some members of Congress and independent analysts have questioned whether the Navy and the Coast Guard need to purchase four different types of small combatants and whether—in spite of the services' well-documented reservations about using similar hull designs—the same type of hull could be employed for certain missions. To explore that possibility, the Congressional Budget Office (CBO) examined three alternatives to the Navy's and the Coast Guard's current plans for acquiring littoral combat ships and deepwater cutters.

■ Option 1 explores the feasibility of having the Coast Guard buy a variant of the Navy's LCS—specifically, the semiplaning monohull—to use as its offshore patrol cutter.

■ Option 2 examines the effects of reducing the number of LCSs the Navy would buy and substituting instead a naval version of the Coast Guard's national security cutter. (The rationale for this option is that, according to some analysts, the NSC's longer mission range and higher endurance might make it better suited than the LCS to act as a "patrol frigate," which would allow the Navy to carry out certain activities—maritime security, engagement, and humanitarian operations—outlined in the sea services' new maritime strategy.)[5]

■ Option 3 examines the advantages and disadvantages of having the Coast Guard buy more national security cutters rather than incur the costs of designing and building a new ship to perform the missions of an offshore patrol cutter.

According to CBO's estimates, all three alternatives and the services' plans would have similar costs, regardless of whether they are calculated in terms of acquisition costs or total life-cycle costs (see Table 1).[6] CBO's analysis also indicates that the three alternative plans would not necessarily be more cost-effective or provide more capability than the services' existing plans. Specifically, even if the options addressed individual problems that the Navy and Coast Guard might confront with their small combatants, it would be at the cost of creating new challenges. For instance, Option 1—which calls for using the LCS monohull for the Coast Guard's OPC—would provide less capability for the Coast Guard from that service's perspective and at a potentially higher cost. Option 2 could provide the Navy with capability that, in some respects, would be superior for executing the peacetime elements of its maritime strategy; but that enhanced peacetime capability would sacrifice wartime capability and survivability. Option 3 would allow the Coast Guard to replace its aging cutters more quickly at a slightly higher cost but without the technical risk that is associated with designing and constructing a new class of ships, which the service's existing plan entails. It would, however, provide fewer mission days at sea and require the Coast Guard to find new home ports for its much larger force of national security cutters.

---

5. U.S. Marine Corps, U.S. Navy, and U.S. Coast Guard, *A Cooperative Strategy for 21st Century Seapower* (October 2007).

6. Acquisition costs are expenses related to developing and buying ships. Total life-cycle costs include acquisition costs as well as costs for operating and replacing the ship over the course of its service life.

Table 1.

Acquisition Costs and Total Life-Cycle Costs of New Surface Combatants Under the Services' Plans and Three Alternative Plans

	Services' Plans		Option 1		Option 2		Option 3	
	Number of Ships	Costs	Number of Ships	Costs	Number of Ships	Costs	Number of Ships	Costs
	Acquisition Costs, 2009 to 2025 (Billions of 2009 dollars)							
LCS	53	33.1 [a]	53	33.2 [a]	28	17.1 [b]	53	33.1 [a]
LCS (Coast Guard variant)	0	0	25	12.1	0	0	0	0
NSC	5	2.9	5	2.9	5	2.6	25	12.5
NSC-Derived Patrol Frigate (Naval variant)	0	0	0	0	20	10.7 [c]	0	0
OPC	25	11.1	0	0	25	11.1	0	0
Total	**83**	**47.1**	**83**	**48.2**	**78**	**41.5**	**78**	**45.6**
Memorandum:								
Average Cost Per Hull (Millions of dollars)[a]	n.a.	570	n.a.	580	n.a.	530	n.a.	580
	Total Life-Cycle Costs, 2009 to 2055[d] (Discounted to net present value)							
LCS	108	65.9	108	65.1	58	35.3	108	65.9
LCS (Coast Guard variant)	0	0	50	23.3	0	0	0	0
NSC	13	10.4	13	10.4	13	9.9	53	31.2
NSC-Derived Patrol Frigate (Naval variant)	0	0	0	0	40	25	0	0
OPC	50	21.6	0	0	50	21.6	0	0
Total	**171**	**97.9**	**171**	**98.8**	**161**	**91.8**	**161**	**97.1**

Source: Congressional Budget Office.

Notes: LCS = littoral combat ship; NSC = national security cutter; OPC = offshore patrol cutter; n.a. = not applicable.

Two littoral combat ships and three national security cutters were purchased prior to 2009.

Total acquisition costs include the costs of developing and buying the ships and the costs of associated LCS mission packages.

Total life-cycle costs include the following: acquisition costs; the costs of replacing each ship one time; and the costs of operating the ships (purchasing fuel, maintaining ship structures and systems, and compensating personnel) between 2009 and 2055.

The Coast Guard's variant of the LCS would be the semiplaning monohull.

a.   Includes $3.4 billion for 62 LCS mission packages.

b.   Includes $1.8 billion for 33 LCS mission packages.

c.   Includes improved antiship missile systems. No LCS mission packages would be purchased for the NSC-derived patrol frigate.

d.   In the calculations of total life-cycle costs presented here, the acquisition costs of the first three NSCs and the first two LCSs are not included because the period that CBO is analyzing begins in 2009. Annual operating costs for all ships are included.

Types of Missions Performed by Small Combatants

The Navy's and the Coast Guard's existing shipbuilding plans reflect the fact that the services have traditionally engaged in distinctly different missions that are performed in different environments. Whereas the Navy's missions have historically been oriented toward combat, the Coast Guard's missions have focused on ensuring maritime security, enforcing maritime law, protecting natural resources, and responding to humanitarian crises in the nation's inland waterways, at its ports, in coastal areas, and on the high seas. What the services' existing plans do not reflect, however, is that the services' new maritime strategy has created some overlap in those mission profiles. Generally speaking, however, even as the services' mission profiles continue to evolve, their respective responsibilities can be characterized as follows:

Coast Guard Missions

The principal missions of the Coast Guard's deepwater cutters require the ships to operate independently with minimal logistical support hundreds, if not thousands, of miles beyond U.S. shores. Traditionally, those missions have centered on activities that are humanitarian in nature or that relate to law enforcement and maritime security. The Coast Guard's mission profile includes the following:

■ Search-and-rescue operations;

■ Environmental disaster response;

■ Fisheries enforcement;

■ Immigration enforcement;

■ Homeland security operations; and

■ Overseas operations in support of deployed military forces.

Those missions often require Coast Guard cutters to engage in solitary at-sea patrols for long periods of time. For example, during particular fishing seasons, the Coast Guard will typically maintain a cutter on-station at the fishing grounds, which may be hundreds of miles from the U.S. coast, to enforce fishing regulations as well as to provide search-and-rescue capability.

Navy Missions

While the Coast Guard's missions for small combatants largely center on peacetime activities, the Navy has designed the LCS-1 and the LCS-2 to perform a variety of wartime and peacetime missions. However, the LCS is above all a warship, the Navy argues, and therefore must be designed and built for combat operations. Specifically, the LCS is designed modularly so that it can be reconfigured fairly quickly to perform one of three distinct missions: finding and sinking quiet diesel-electric submarines operating in crowded, noisy, and shallow coastal waters; finding and neutralizing mines; and countering swarm attacks by small, high-speed boats armed with missiles. The ships displace about 3,300 tons fully loaded and can cruise at speeds in excess of 40 knots.

At the same time, the sea services' new maritime strategy places about equal emphasis on peacetime presence and engagement missions: maritime security, building naval partnerships with the navies of other countries, and humanitarian operations.[7] These are peacetime missions that resemble certain Coast Guard missions, but the Navy would generally conduct its peacetime missions overseas.

While almost any Navy ship could be called on to participate in missions requiring "low-end" capability—maritime security operations, for instance, which range from antipiracy and counterdrug operations to enforcing economic sanctions—small surface combatants are especially useful in that capacity because they have sufficient capability to perform such missions but are small enough to operate in shallower waters, where those activities often take place. The Navy would prefer not to use its large surface combatants for those operations, for several reasons: The vessels were designed and built for missions requiring "high-end" capability, such as fleet air defense and land attack; they are not designed to operate quickly and efficiently in shallower waters; and small surface combatants are capable of carrying out low-end missions but at a fraction of the cost of a large surface combatant. In particular, although small surface combatants can carry helicopters and small boats to perform interception and boarding operations, a ship with long range and high endurance would provide advantages for executing maritime security operations because those ships might be required to patrol large areas over extended periods of time.

---

7. *A Cooperative Strategy for 21st Century Seapower*, p. 11.

Similarly, small combatants are effective platforms for engaging with the navies and coast guards of other, smaller nations. The maritime forces of most countries that the U.S. Navy encounters are composed of small ships (few are larger than 3,000 to 4,000 tons). Thus, when considering engagement or partnership activities (including personnel exchanges and joint training exercises), the maritime forces of other countries might find it difficult to participate if the U.S. Navy had only large surface combatants or even larger amphibious ships with which to conduct those activities. The availability of small ships could thus enhance the Navy's ability to maintain and expand its engagement and mentoring activities with the navies of other nations.

Along with larger surface combatants, small combatants are also useful in conducting humanitarian operations. They can sail closer to shore than larger ships, extending the reach of the Navy's task forces both physically and with their embarked helicopters. Small combatants would contribute to delivering supplies, rescuing the stranded and injured, or policing the seas to maintain order in the aftermath of a natural disaster or some other type of humanitarian crisis.

This mission profile for small combatants raises a question: The Navy determined its key design characteristics for the LCS before the sea services formulated, wrote, and promulgated the new maritime strategy emphasizing maritime security operations, engagement and partnership with the navies of other countries, and humanitarian response. To what extent should that emphasis and those activities affect the design of the littoral combat ships or the composition of the Navy's and Coast Guard's overall small combatant force?

The Navy's and the Coast Guard's Current Small Combatants

The Navy and the Coast Guard today deploy a total of 83 small combatants and mine-warfare ships. The Navy's current inventory of small combatants includes two classes of ships:

■ 30 Oliver Hazard Perry class FFG-7 guided-missile frigates; and

■ 14 Avenger class MCM-1 mine-countermeasures ships.

The Coast Guard's current inventory of cutters, by ship size and class, are as follows:

■ 12 Hamilton class 378-foot high-endurance cutters;

■ 13 Bear (also known as Famous) class 270-foot medium-endurance cutters; and

■ 14 Reliance class 210-foot medium-endurance cutters.

The characteristics and capabilities of the Navy's and Coast Guard's existing small combatants are described below.

Oliver Hazard Perry Class Guided-Missile Frigate

Oliver Hazard Perry class guided-missile frigates, or FFG-7s, are the Navy's smallest surface combatants, displacing about 4,100 tons each. Originally designed as escorts for resupply convoys that would cross the Atlantic in the event of a war with the Soviet Union, the frigates have been adapted to post–Cold War uses, including maritime security operations (such as sanctions enforcement or counterdrug operations). The ships can carry two H-60 type helicopters and, because they are equipped with towed array sonar, can conduct antisubmarine warfare operations. According to one source: "The soundness of the design has permitted the expansion [of capabilities], and the ships have proven remarkably sturdy."[8] The average cost of the 51 ships purchased by the U.S. Navy between 1973 and 1984 was about $570 million each (in 2009 dollars), but later ships, which featured improved antisubmarine capabilities, cost about $650 million each.

At an average age of 25 years, however, ships in this class are reaching the end of their projected 30-year service life (see Figure 1). In recent years, in an effort to save money, the Navy has removed much of the ships' armament; consequently, FFG-7s no longer have the capability to launch antiship or surface-to-air antiaircraft missiles despite their designation as guided-missile frigates.[9] (FFG-7s still retain certain combat systems, including open-ocean

8. A.D. Baker III, *The Naval Institute Guide to Combat Fleets of the World, 2002–2003: Their Ships, Aircraft, and Systems* (Annapolis, Md.: Naval Institute Press, 2002), p. 965.

9. Prior to the removal of this armament, the FFGs would typically carry 36 antiaircraft missiles and 4 Harpoon antiship missiles.

Figure 1.

Average Age of the Navy's and the Coast Guard's Existing Small Combatants

(Years)

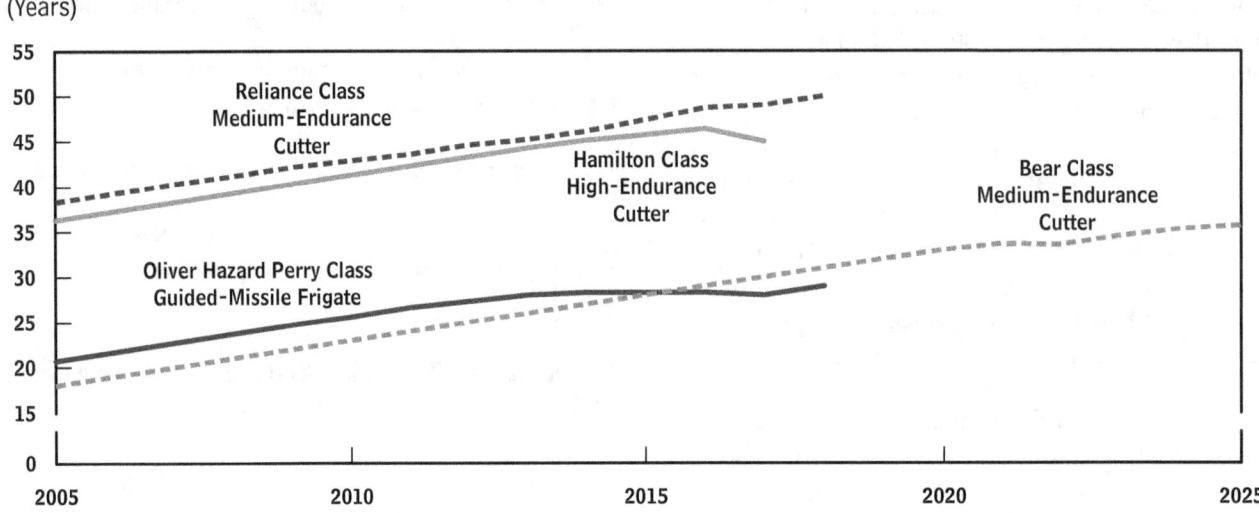

Source: Congressional Budget Office based on data from the U.S. Navy and the U.S. Coast Guard.

Note: The average age of ships in each class is displayed through the projected year of retirement.

antisubmarine capability.) Nor does the Navy consider the Perry class suitable for the type of littoral operations against mines, diesel-electric submarines, and fast small boats for which the LCS is being built. For instance, Perry class ships do not have the space to carry and operate the off-board unmanned systems that are at the heart of the Navy's strategy to defeat such threats in the future.

Avenger MCM-1 Mine-Countermeasures Ship
The 14 Avenger class mine-warfare ships in the fleet each displace about 1,300 tons and represent the only dedicated antimine craft in the U.S. Navy. The first ship was commissioned in 1987, the last in 1994; they are slated to retire between 2018 and 2026, with their mission being assumed by the LCS's mine-warfare package. (Earlier shipbuilding plans, however, envisioned replacing the Avengers with a dedicated mine-clearing ship as well as purchasing 55 LCSs with more than the 24 mine-warfare packages than are currently programmed.)[10]

The Avenger class ships were built with a fiberglass-sheathed wood hull to keep the vessel's magnetic signature as low as possible to resist detection by magnetic mines. They have less than half the displacement and are

less than half the length of the LCS and can support a crew of 83. The Avenger program also suffered a number of design and construction problems that increased the cost of ships in that class. In 2009 dollars, the average cost of the 14 ships was about $260 million each.

Originally intended to protect U.S. ballistic missile submarines against deep-ocean Soviet mines, ships in the Avenger class have been fitted with various equipment upgrades to allow them to detect mines in both shallow and deep water. During the first Gulf War, the *Avenger* was deployed for operations in the Persian Gulf, and four Avenger class minesweepers participated in countermine operations during Operation Iraqi Freedom. Today four remain home-ported in Bahrain.

Hamilton Class High-Endurance Cutter
The mainstay of the Coast Guard's existing Deepwater assets for the past 40 years, these 378-foot ships displace about 3,100 tons and have a range of 14,000 nautical miles at 11 knots. Originally, they were built not only to perform the long-range missions and patrols sometimes required of Coast Guard ships but also with the capability to perform some combat-oriented operations in the event the Cold War led to overt hostilities. For example, during the Cold War, the ships were equipped with antiship missiles and antisubmarine weapons, including torpedo

---

10. Department of the Navy, *A Report to Congress on Annual Long-Range Plan for the Construction of Naval Vessels* (May 2003 and May 2004).

tubes, in order to protect U.S. convoys sailing for Europe to reinforce NATO forces. Those weapons were ordered removed by the Coast Guard commandant in 1992.

At an average age of 40 years, however, the ships are reaching the end of their useful service life. A recent report from *Inside the Navy* states that the Coast Guard is assessing the readiness of the entire class after the discovery of "immense structural problems" in two of the ships.[11] Those problems included corrosion in the hull, which compromised the structural integrity of the vessels. Nevertheless, the Coast Guard expects to continue operating ships in the Hamilton class until the new NSCs are available to replace them.

Bear Class Medium-Endurance Cutter
These 270-foot cutters are the Coast Guard's newest and most sophisticated ships. Built mostly in the 1980s, ships in this class average 22 years in age. They displace 1,800 tons, can reach a maximum speed of about 20 knots, and have a mission range of 9,900 nautical miles when cruising at 12 knots. They were designed to be able to conduct a 14-day patrol at a range of 400 nautical miles. The cutters can "embark" (that is, carry, operate, and sustain) one search-and-rescue helicopter from the H-60 or H-65 family. Nevertheless, Bear class cutters have been criticized for their slow speed and for the fact that they do not move smoothly in heavy seas, which has made it difficult to perform certain Coast Guard missions.[12]

If the Coast Guard executes its Deepwater plan according to its most recent schedule, which calls for purchases of OPCs to begin in 2015, the Bear class cutters would not serve as long as the 210-foot Reliance or 378-foot Hamilton class cutters. However, they would exceed their notional retirement age of 28 years when the last of them leave the fleet in 2025.

Reliance Class Medium-Endurance Cutter
Although built in the 1960s, these 210-foot cutters were extensively modernized in the 1990s. As with all of the Coast Guard's large cutters, the lack of an available replacement has kept the ships operational longer than the service would prefer. Originally designed to operate

for 30 years, the average age of ships in the class is currently 42 years. Reliance class cutters displace about 1,000 tons, can reach a maximum speed of 18 knots, and have a range of 6,100 nautical miles when cruising at 13 knots (see Table 2). Unlike larger Coast Guard cutters, these ships cannot embark a helicopter, although they do have a landing deck that can temporarily support helicopter operations. The ships were designed primarily for the purpose of performing law-enforcement and search-and-rescue operations.

A medium-endurance cutter is one that can engage in a three-week patrol without the need for replenishment of stores. The normal patrol for one of these ships is about six to seven weeks, during which the vessel returns to port for refueling and replenishment once. A medium-endurance cutter would typically spend about half the year at sea and the other half in home port to allow the ship to undergo mechanical and structural maintenance and the crew to rest and receive additional training.

The Navy's and the Coast Guard's Future Small Combatants
Both services are pursuing programs to replace their existing small combatants as those ships reach the end of their service life. The Navy plans to replace its Oliver Hazard Perry class guided-missile frigates and Avenger class mine-countermeasures ships with 55 Freedom class littoral combat ships (although the exact mix of the two variants has not been decided upon).[13] The Coast Guard will replace its aging Hamilton, Bear, and Reliance class cutters with 8 Legend class national security cutters and 25 offshore patrol cutters.[14] The status of those new programs is discussed below.

Freedom Class Littoral Combat Ship
The Navy's 2009 shipbuilding plan details the service's intention to build 53 littoral combat ships between 2009 and 2019. The Navy ordered the first two Freedom class

---

11. Rebekah Gordon, "Coast Guard to Assess Readiness of All High-Endurance Cutters," *Inside the Navy* (December 8, 2008).

12. Norman Polmar, *The Naval Institute Guide to Ships and Aircraft of the U.S. Fleet,* 18th ed. (Annapolis, Md.: Naval Institute Press, 2005), p. 581.

---

13. When the Navy initiated the LCS program in November 2001, it also operated 12 Osprey class MHC-56 coastal minehunters. In essence, the Navy is replacing 56 small combatants (30 FFG-7s, 14 MCMs, and 12 MHCs) with 55 LCSs.

14. The Coast Guard is replacing 39 ships with 33 NSCs and OPCs. It is currently reviewing its requirements for the national security and offshore patrol cutters. At the time this report was written, the Coast Guard had not announced a change in the number of ships it plans to purchase but that still had to be determined.

Table 2.

Characteristics and Capabilities of the Navy's and the Coast Guard's Existing Small Combatants

	Navy	Coast Guard		
	(Oliver Hazard Perry Class Guided-Missile Frigate)	(Hamilton Class High-Endurance Cutter)	(Bear Class Medium-Endurance Cutter)	(Reliance Class Medium-Endurance Cutter)
Length (Feet)	455	378	270	210
Beam (Feet)	45	43	38	34
Draft (Feet)	22	20	14	11
Full-Load Displacement (Long tons)	4,100	3,100	1,800	1,000
Maximum Speed (Knots)	29	29	20	18
Endurance (Days)	n.a.[a]	45	21	21
Range (Nautical miles)	5,000[b]	14,000[c]	9,900[d]	6,100[e]
Number of Helicopters	2	1	1	0
Design Service Life (Years)	30	30	28	30
Current Average Age (Years)	25	40	22	42
Crew	239	178	116	75

Source: Norman Polmar, *The Naval Institute Guide to Ships and Aircraft of the U.S. Fleet*, 18th ed. (Annapolis, Md.: Naval Institute Press, 2005).

Notes: n.a. = not applicable.

Beam indicates the width of the ship. Draft indicates the depth to which the ship is immersed. Full-load displacement includes the weight of the ship plus its crew, cargo, weapons, and fuel.

a.   The Navy does not consider endurance to be a meaningful measure of capability because of the availability of logistics ships that operate with Navy combat ships to provide replenishment of stores and refueling at sea.

b.   At 18 knots.

c.   At 11 knots.

d.   At 12 knots.

e.   At 13 knots.

LCSs in 2005 and 2006, respectively. Thus, the Navy plans to purchase a total of 55 of those ships. Currently the Navy is building two different designs, one a semi-planing monohull (the ship ordered in 2005; see Figure 2) and the other a trimaran (the ship ordered in 2006). The Navy has not yet determined when or if it will eventually select one hull or whether it will divide the entire class between the two types of ship.

The LCS differs from the Navy's existing and previous warships in that the program is divided into two components: the sea frame (the ship itself) and mission packages (combat systems). The sea frame will be built with the ability to switch mission packages, depending on which mission the ship is intended to carry out at a given time: antiship warfare, antisubmarine warfare, and counter-mine warfare. The Navy expects to buy 64 of these mission packages for the 55-ship program—24 for antiship warfare, 16 for antisubmarine warfare, and 24 for anti-mine warfare. Once it has more familiarity with the ship, the Navy may develop and then procure other mission packages for other types of missions.

Bringing the LCS program to fruition has been difficult. The Navy originally hoped that each sea frame would

Figure 2.

Line Drawing of the Navy's Freedom Class Littoral Combat Ship

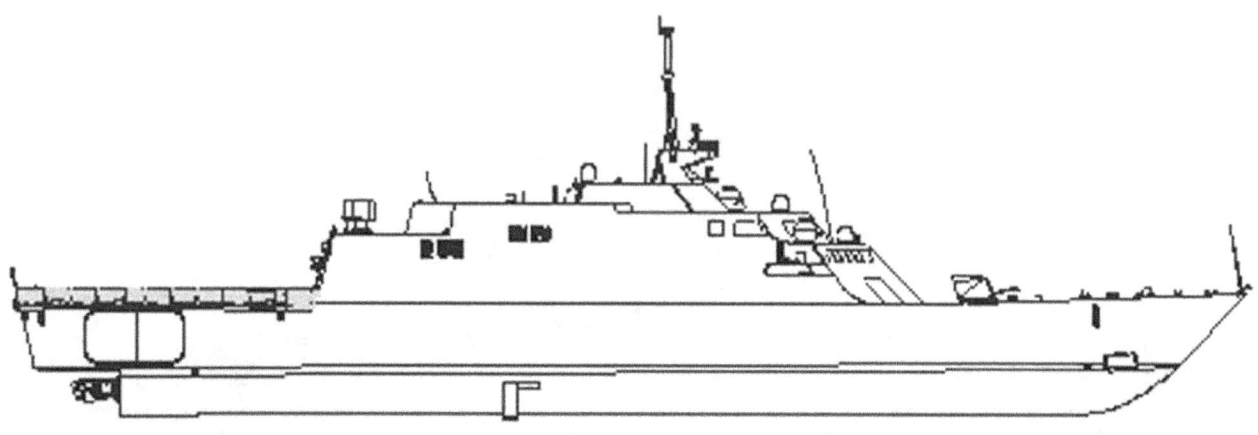

Source: Lockheed Martin Corporation.

Note: This illustration depicts the semiplaning monohull variant of the littoral combat ship.

cost about $260 million in 2009 dollars and take two years to build. The first two ships, however, are now expected to cost about $700 million each; CBO estimates that the average cost of subsequent ships will be about $550 million and that they will take about three years to build. The average cost of each mission package would add about $60 million to that figure.

Legend Class High-Endurance Cutter

The Coast Guard's new high-endurance cutter—also known as the national security cutter—is designed to replace the 378-foot Hamilton class cutters. Ships in this class displace about 4,300 tons and can accommodate a crew of 120, with the capacity to berth as many as 150 people. The ship, which is about the same size as the Navy's Oliver Hazard Perry class frigate, has a range of about 12,000 nautical miles when steaming at 8 to 9 knots. It is designed to perform the Coast Guard's most challenging missions, such as patrols far from U.S. shore, and can operate with U.S. Navy fleet units on missions in other parts of the world.

The first ship of the class, the *Bertholf* (see Figure 3), has been delivered to the Coast Guard, but the program has experienced significant cost increases and schedule delays. When construction of the ship began in 2004, the Coast Guard expected it would join its fleet in 2006, but the ship was not delivered until 2008. Originally projected to cost $475 million, the ship's costs have increased to about

$750 million. Those cost increases have affected production of subsequent ships of the class by increasing the overall cost of the program by about 40 percent.[15]

Notional Medium-Endurance Cutter

Under the Coast Guard's original Deepwater plan, the service's new medium-endurance cutter, often referred to as the offshore patrol cutter, would likely have been a scaled-down version of the national security cutter. In fact, senior officials in the Coast Guard and industry officials both argued that the government would get the best price for the OPC by maximizing the commonality between the larger and smaller cutters. However, in the wake of the troubles with the national security cutter and, in particular, the cost increases the program has experienced, the Coast Guard has taken over direct management of the Deepwater program and has opened up construction of the offshore patrol cutter to general competition. Under the original plan, the first OPC was to be bought in 2010; the Coast Guard now expects to award the first OPC contract in 2014 or 2015.

The service recently released a request for information (RFI) outlining its design criteria for the OPC. Those

---

15. For an extensive discussion of these and other problems, see Ronald O'Rourke, *Coast Guard Deepwater Acquisition Programs: Background, Oversight Issues, and Options for Congress,* CRS Report for Congress RL33753 (October 9, 2008).

Figure 3.

Line Drawing of the Coast Guard's Legend Class High-Endurance Cutter

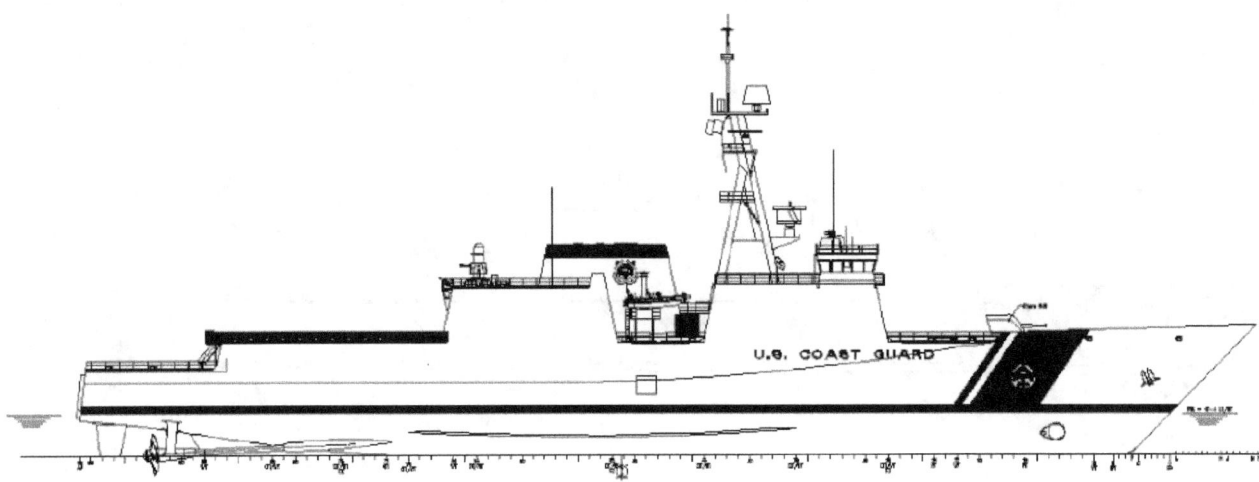

Source: Northrop Grumman Corporation.

Note: The high-endurance cutter is also referred to as the national security cutter.

requirements include the ability to a achieve a sustained speed of 25 knots; the capacity to berth 120 crew members; the capability to reach a range of 7,500 nautical miles at 12 to 14 knots with a fuel reserve of 30 percent (the NSC's 12,000 mile range does not include a reserve); and the ability to launch and recover small boats in sea state five (characterized as rough seas with wind speeds of 21 to 25 knots and waves 8 to 12 feet high), conduct more than one small boat operation at the same time, and embark one helicopter. A critical requirement in the RFI is that the proposed OPC should have a parent craft (an existing ship on which the new ship's design is based) that is constructed to the American Bureau of Shipping's Naval Vessel Rules (NVR).[16] It is not clear how many viable candidates there are that can meet this particular requirement because many such parent craft are foreign-built and, thus, not subject to the NVR.

The Coast Guard also stated, however, that it will consider ship candidates that do not meet these requirements exactly, implying that the service will weigh the advantages of various capabilities and costs against one another. In particular, the RFI stated that an OPC proposed by industry could have a range of between 5,500 and 9,000 nautical miles, endurance of 30 to 50 days, and accommodations for 90 to 130 people. With respect to the

NVR requirement, the Navy has waived areas of non-compliance with the NVR in the past when it felt the need to do so; the Coast Guard could as well.

For the purpose of this analysis, CBO assumed the Coast Guard would acquire an OPC that conforms to the criteria outlined in the RFI and that the ship would displace about 3,700 tons when fully loaded. That displacement was based on the notional displacement of the OPC under the original Deepwater plan but including post–September 11 Homeland Security requirements.

Key Characteristics of the Services' Future Small Combatants

While a particular ship design comprises many characteristics, four appear particularly important to the Coast Guard: operating environment, endurance, the ability to operate in austere ports, and berthing capacity. The Coast Guard's deepwater cutters must be prepared to operate, without logistical support, in waters far from U.S. ports and in environments where the sea state can be level 5 or greater. For example, the Coast Guard routinely operates ships in the Bering Sea off Alaska for search-and-rescue operations, fisheries enforcement, and environmental protection. Coast Guard deepwater cutters thus are designed to allow long cruising ranges and high endurance. Those needs stem from the lack of an extensive

---

16. The next section discusses the Naval Vessel Rules in more detail.

Coast Guard resupply capability, such as that found with the Navy's fleet of combat logistics ships (including oilers and multiple-product vessels).

Coast Guard operations along much of the U.S. shoreline may involve putting into ports that do not have many services or facilities to assist in the docking of ships. Tug service, for example, may not be available. To compensate, bow thrusters, which help a ship dock on its own, have been a design feature of Coast Guard cutters for decades.

Finally, the Coast Guard's various law enforcement and immigration responsibilities make it desirable to have berthing capacity well in excess of that needed to house the crew. The service's various missions require that cutters periodically carry other law enforcement personnel (from the Federal Bureau of Investigation or the Drug Enforcement Agency, for instance) or have the capability to house immigrants who might be picked up at sea when attempting to the enter the United States illegally.

In contrast with the Coast Guard's cutters, the LCS is optimized for high-speed, shallow-water combat operations with a minimal crew. The LCS is designed for sprint speed of greater than 40 knots compared with 28 knots for the NSC and, prospectively, 25 knots for the OPC. The LCS's speed and shallow-water capability are partially a function of the fact that the vessel's draft is 6 feet less than that of the NSC but only slightly less than the 16 feet desired for the OPC.

A critical difference between the Coast Guard's cutters and the Navy's littoral combat ship is the level of survivability for which the Navy's vessels are designed and constructed. Because the LCS is considered primarily a warship, the Navy has designed it to sustain some degree of damage during combat (the exact amount is classified) and still remain afloat. Important elements of that survivability include stronger bulkheads, watertight compartmentalization, and extensive damage-control systems. Such higher survivability standards add to a ship's costs, which the Coast Guard has been generally unwilling to pay for in its cutter programs. The reason is straightforward: Although the service may expect its cutters to be called to serve in wartime operations, those vessels are not expected to participate in direct combat operations. For example, during the initial phases of Operations Desert Storm and Iraqi Freedom, Coast Guard cutters were not involved with destroying Iraq's small naval forces or clear-

ing mines from ports. Once those early activities were concluded, however, Coast Guard cutters operated (and continue to operate) in the Middle East, providing patrols and contributing to maritime interception activities. The fact that the Navy's ships and the Coast Guard's cutters are built to different levels of survivability does not mean that the Navy could not adopt a Coast Guard cutter as a naval ship platform; if it did so, however, it would be taking on more risk than it would prefer if the ship was used in combat.

Another issue is that Naval Vessel Rules are being adhered to in the construction of the Navy's LCS but not for the construction of the Coast Guard's NSC. Naval Vessel Rules are a set of standards, specifications, and requirements developed by the American Bureau of Shipping in conjunction with the Navy to guide the construction of naval ships. They are analogous to the building codes used to construct commercial or residential buildings. The type of ship one plans to build will determine which provisions of the NVR are relevant to that particular ship program. The national security cutter was not built to NVR standards because those guidelines did not yet exist when the first NSCs were designed and built. The LCS program was also under way when the NVR came into existence, but the Navy decided to apply the rules to the LCS retroactively—a decision that caused the costs of the LCS to go up. (The Navy and the LCS's shipbuilders disagree about exactly how much of the ship's cost growth is attributable to the adoption of NVR.) The Coast Guard has stated that it expects to use NVR in the design and construction of its offshore patrol cutter, even though the OPC is not a warship. The Coast Guard would employ only those parts of NVR that would be relevant to the type of cutter the service plans to buy. Thus, both the littoral combat ship and the offshore patrol cutter will be built according to the Naval Vessel Rules, but the LCS will still be constructed as a warship and the OPC will not.

Projected Costs of the Services' Small Combatant Programs Under Current Acquisition Plans

Over the next 10 years, the Navy plans to build its new class of littoral combat ships, and the Coast Guard plans to replace its fleet of aging cutters with a new generation of high- and medium-endurance cutters. According to

Figure 4.

The Navy's and the Coast Guard's Purchases and Inventory of Small Combatants

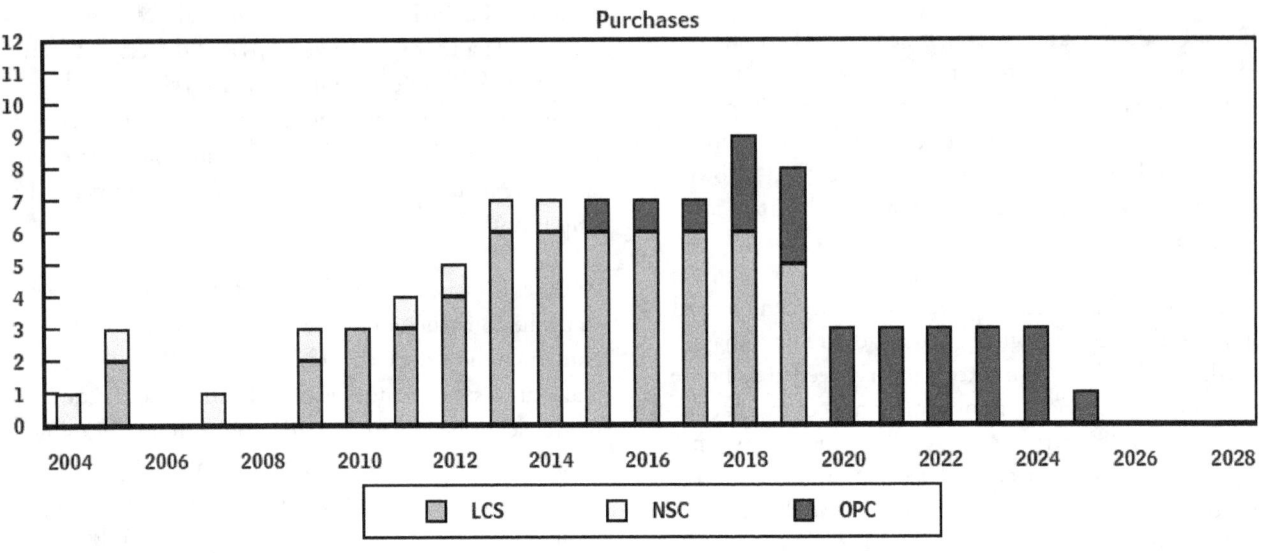

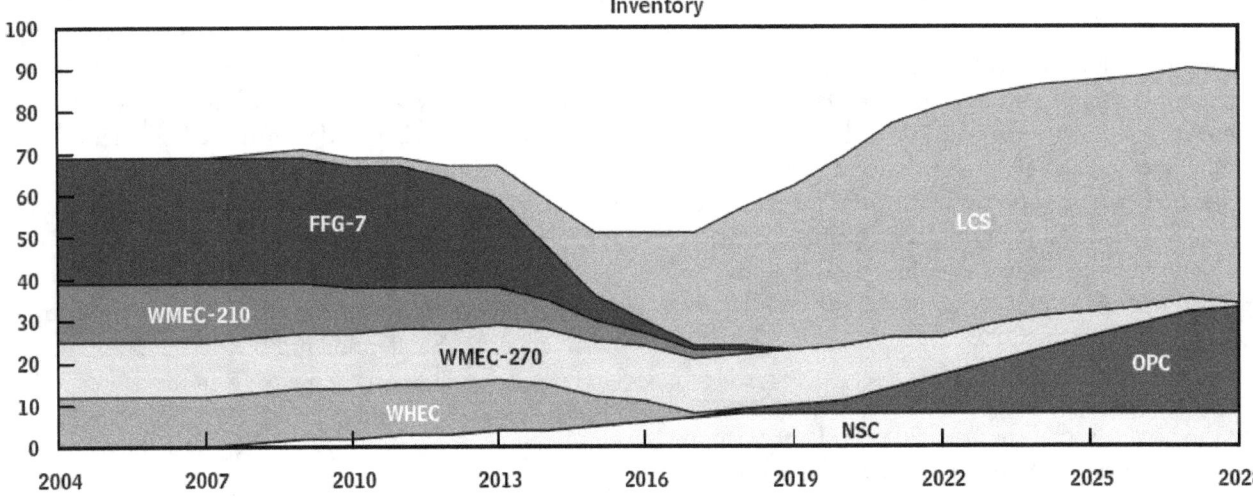

Source: Congressional Budget Office based on data from the U.S. Navy and the U.S. Coast Guard.

Notes: FFG-7 = Oliver Hazard Perry class guided-missile frigate; LCS = Freedom class littoral combat ship; NSC = Legend class national
security cutter; OPC = offshore patrol cutter; WHEC = Hamilton class high-endurance cutter; WMEC-210 = Reliance class 210-foot
medium-endurance cutter; WMEC-270 = Bear class 270-foot medium-endurance cutter.

Purchases of new combatants are based on the latest information from the services, although the procurement schedules are
subject to change.

the Navy's 2009 budget submission, the service plans to
purchase a total of 55 LCSs by 2019, ramping up to a
construction rate of six ships per year (see the top panel of
Figure 4). The Navy has not yet determined what portion
of its fleet will consist of semiplaning monohulls and
what portion will consist of trimarans. Thus, for the purpose of this analysis, CBO assumed that the Navy would
purchase 28 monohull and 27 trimaran LCSs and in

roughly equal annual proportions over the course of the
program.

Under its current acquisition schedule, the Coast Guard
would purchase one national security cutter per year
between 2011 and 2014, for a total of eight. The first
four, including one purchased in 2009, have already been
ordered. Once the last NSC is ordered, the service will
begin purchasing 25 OPCs. The first ship would likely be

Figure 5.

Acquisition Costs of the Services' Small Combatant Programs Under Existing Plans, 2009 to 2025

(Billions of 2009 dollars)

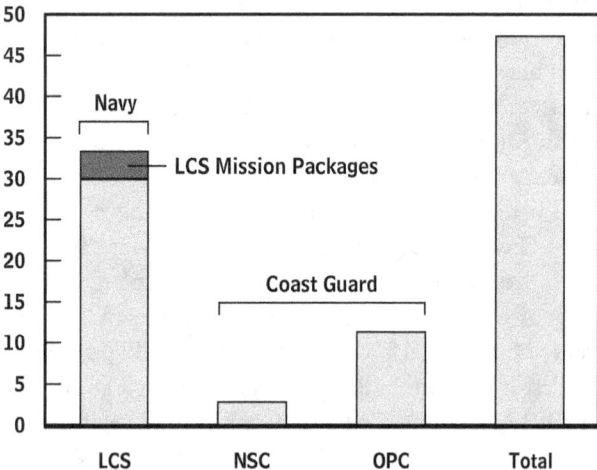

Source: Congressional Budget Office.

Notes: LCS = littoral combat ship; NSC = national security cutter; OPC = offshore patrol cutter.

Total acquisition costs include the costs of developing and buying the ships and the costs of associated mission packages.

The LCS can be equipped with one of three mission packages (combat systems), depending on the type of mission the ship is intended to carry out at a given time: antiship warfare, antisubmarine warfare, or countermine warfare.

purchased in 2015, followed by one more each in 2016 and 2017. Production would then ramp up to three per year in 2018, with the final ship purchased in 2025.

As new deepwater cutters enter the Coast Guard's fleet, the service will begin retiring its aging high- and medium-endurance cutters. Under the notional retirement schedule the Coast Guard is using for planning purposes, the last Hamilton class cutters would retire in 2017, the last Bear class cutters would retire in 2028, and the last Reliance class ship would retire in 2018 (see the bottom panel of Figure 4). The retirement schedule could be adjusted, depending on perceived operational needs as well as the expected delivery dates of new cutters.

The relationship between the Navy's procurement of new LCSs and the retirement of its frigates is more tenuous. In the late 1990s and early 2000s, the Navy originally

intended to replace the Perry class frigates and Spruance class destroyers with 32 new DD-21 destroyers. That plan envisioned no role for frigates in the future Navy. In 2001, the service launched the LCS program, which would in essence replace 30 FFGs and 26 mine-warfare ships with 55 littoral combat ships. Under earlier plans, the LCSs would have been completed as the last of the frigates were retired from service but before the last 14 mine-warfare ships were retired. Under the Navy's current retirement schedule, the Perry class frigates will retire by 2019, having reached the end of their service life, while the LCSs will enter the fleet as fast as the program's progression and budgets will allow. All mine-warfare ships would be retired by 2026.

Overall, the Navy and the Coast Guard plan to spend about $47 billion purchasing small combatants over the 2009–2025 period, CBO estimates (see Figure 5). The Navy's LCS program represents the largest share of that amount, or about $33 billion for the ships and mission packages. CBO estimates that the average cost of an LCS would be about $550 million, with another $60 million or so on average required for each mission package. The Coast Guard will spend about $3 billion buying five more NSCs and an additional $11 billion on 25 OPCs. The average per-ship cost of the Coast Guard's NSCs and OPCs would be about $580 million and $450 million, respectively. CBO estimated the cost of the OPC using the cost per ton of the NSC and then adjusting for the smaller displacement that the OPC would likely have. Over the course of the next 10 years, the Navy and Coast Guard combined will spend an average of about $3 billion per year on small combatants.

Alternatives to the Services' Current Acquisition Plans

Because of the budgetary pressures currently facing all of the armed services, some members of Congress and independent analysts have questioned whether the Navy and the Coast Guard need four types of small combatants—two versions of the LCS for the Navy; and two cutters, the NSC and the OPC, for the Coast Guard—and whether the services could employ the same ship or hull for certain missions. To explore the implications of combining or otherwise modifying the services' small combatant programs, the Congressional Budget Office examined the potential advantages and disadvantages of three alternatives to the Navy's and the Coast Guard's current acquisition plans. The first option explores the feasibility

Figure 6.

Total Acquisition Costs of the Services' Small Combatant Programs Under Three Alternative Plans, 2009 to 2025

(Billions of 2009 dollars)

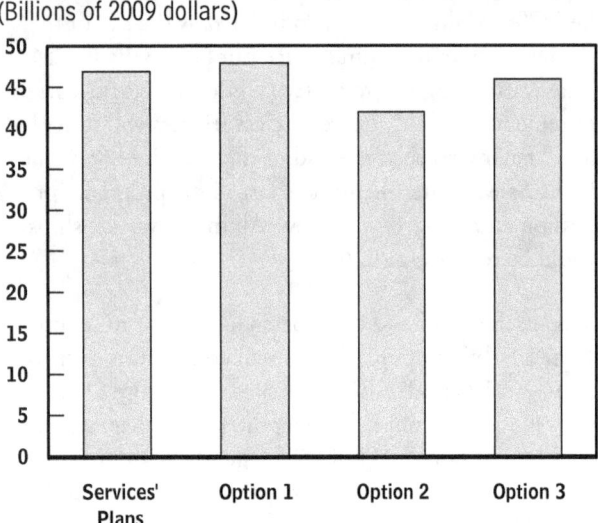

Source: Congressional Budget Office.

Notes: Total acquisition costs include the costs of developing and buying the ships and the costs of associated mission packages.

Under the three alternative plans, the services' existing plans for their small combatant programs would be modified as follows:

- Option 1: The Coast Guard would buy a variant of the Navy's littoral combat ship (LCS)—specifically, the semiplaning monohull—to use as its offshore patrol cutter (OPC). Between 2009 and 2025, the Navy and Coast Guard would purchase a total of 78 LCSs.

- Option 2: The Navy would purchase only 30 LCSs—as opposed to the 55 currently planned—and buy 20 variants of the Coast Guard's national security cutter (NSC) to use as patrol frigates.

- Option 3: The Coast Guard would cancel its OPC program and buy an additional 20 NSCs (for a total of 28) instead.

of having the Coast Guard buy a variant of the LCS monohull to use as its offshore patrol cutter. The second option looks at reducing the number of LCSs the Navy would buy and substituting instead a naval version of the national security cutter as a patrol frigate, which may be better suited than the LCS to carry out nonwarfighting elements of the sea services' maritime strategy. The third option examines the pros and cons of having the Coast Guard buy more national security cutters rather than

design and build a new ship to perform the missions of the offshore patrol cutter.

Option 1: Base the Coast Guard's Medium-Endurance Cutter on a Variant of the Navy's LCS Monohull

Under this option, the Navy and the Coast Guard would use a common hull in their small combatant programs. As its new OPC, the Coast Guard would purchase a slightly modified version of the LCS that incorporated additional fuel tanks and berthing racks and had a reduced maximum speed to better meet the Coast Guard's requirements for endurance (see Table 3). The Coast Guard would purchase the LCS-OPC at the same rate called for under its original Deepwater plan, which CBO used as the basis for its analysis of the service's program. Thus, in 2015, the Coast Guard would purchase the first LCS monohull, followed by one each in 2016 and 2017. In 2018, the Coast Guard would begin purchasing three per year until the total purchase of 25 was complete in 2025. The acquisition schedule under this option would thus remain the same as that called for under the Coast Guard's current plans. The rate at which the Navy acquired LCSs would slow slightly, however. Under this option, the two services combined would not purchase more than six ships per year, and the Navy would not complete the purchase of 55 LCSs until 2022 (as opposed to 2019 under its 2009 shipbuilding plan. CBO chose to slow the purchase of the Navy's LCSs rather than delay the start of the OPC because of the advanced aged of the Coast Guard's existing ships). Between 2009 and 2025, the Navy and Coast Guard would purchase a total of 78 littoral combat ships, 35 of which would be built at one yard and 43 of which would be built at a second yard. The precise split, if any, between the two types of LCS that the Navy would purchase (monohull or trimaran) would be determined later; but the Coast Guard would build 25 LCS monohulls that were slightly modified for the service's missions. The Coast Guard would also purchase the remaining five national security cutters in its program.

The overall cost to the government under this option would be somewhat higher than under the services' plans (see Figure 6). The combined cost of the LCS and NSC programs for both services would total $48.2 billion through 2025 versus $47.1 billion for the services' plans. The average cost for all of the Coast Guard's and Navy's

Table 3.

Characteristics and Capabilities of the Coast Guard's New and Notional Small Combatants

	NSC (Legend Class)	OPC (Requirement)[a]	OPC (Range of Acceptable Submissions)	LCS (Coast Guard Variant)[b]
Length (Feet)	418	N.A.	300 to 390	378
Beam (Feet)	54	N.A.	N.A.	57
Draft (Feet)	22	N.A.	up to 18	14
Full-Load Displacement (Long tons)	4,300	~3,700[c]	N.A.	3,500
Maximum Speed (Knots)	28	25	30	~30
Endurance (Days)	60	45	30 to 50	21[d]
Range (Nautical miles)	12,000	7,500[e]	5,500 to 9,000	6,300[e]
Operating Environment	SS5	SS5	SS5	SS4
Number of Helicopters	2	1	1	2
Constructed to Naval Vessel Rules	No	Yes	Yes	Yes
Bow Thruster	Yes	Yes	N.A.	No
Service Life (Years)	30	30	25 to 40	25
Number of Berthing Racks	148	120	90 to 130	120[f]

Source: Congressional Budget Office.

Notes: NSC = national security cutter; OPC = offshore patrol cutter; LCS = littoral combat ship; N.A. = not available; SS4 = sea state 4; SS5 = sea state 5.

Beam indicates the width of the ship. Draft indicates the depth to which the ship is immersed. Full-load displacement includes the weight of the ship plus its crew, cargo, weapons, and fuel.

Sea state refers to the condition of the seas, such as wind speed and wave height. The higher the sea state the rougher the seas. Sea state 4 is characterized by winds of 18 to 20 knots and waves of 6 to 7.5 feet. Sea state 5 is characterized by winds of 21 to 25 knots and waves of 8 to 12 feet.

a. Specifications for the OPC are based on a request for information that the Coast Guard submitted to the industry in October 2008 when soliciting designs for a new ship.

b. This comparison is based on the assumption that the Coast Guard would use the semiplaning monohull variant of the Navy's LCS.

c. The ship's displacement is based on notional design specifications proposed for the OPC under the Coast Guard's Deepwater plan.

d. According to industry officials, endurance could be expanded either by airlifting supplies to the ship or by adding more refrigeration to the mission-package spaces on the LCS.

e. Assumes a fuel reserve of 30 percent.

f. Expanding the number of berths from 100 to 120 would require using space in the mission bay.

Figure 7.

Number of Mission Days Provided Annually by the Coast Guard's Total Inventory of Cutters

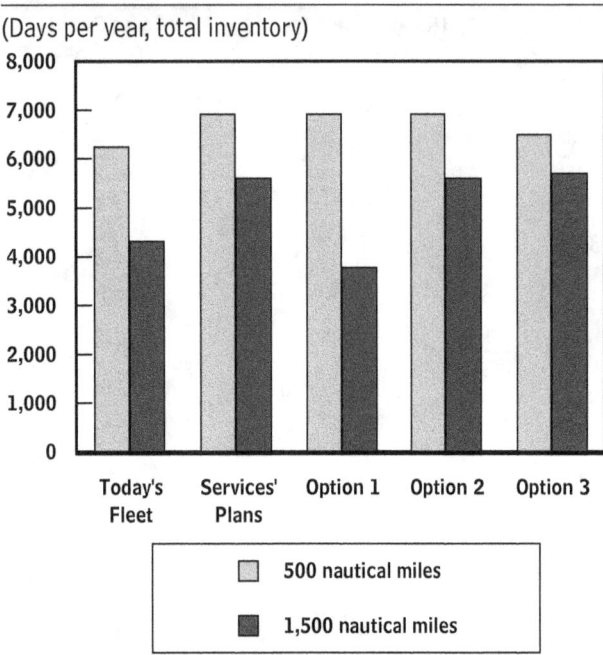

(Days per year, total inventory)

Legend:
- 500 nautical miles
- 1,500 nautical miles

Source: Congressional Budget Office based on data from the U.S. Navy and the U.S. Coast Guard.

small combatants would be about $580 million per ship, including the Navy-specific mission-warfare packages for the LCSs, compared with $570 million per ship under the services' plan. The total life-cycle costs of Option 1 would be $99 billion in discounted dollars compared with $98 billion under the services' plans. (See Box 1 for an explanation of how CBO calculated total life-cycle costs.)

Advantages of Option 1. Pursuing a common hull for the Navy's and the Coast Guard's small combatants would provide a larger force of small combatants that could conduct operations in conditions where the ships might come under enemy fire. Eighty ships would fall within that category—55 Navy LCSs and 25 Coast Guard OPCs—compared with only 55 Navy LCSs under the services' plans. In effect, the more survivable LCS hull would enable the Coast Guard to serve as a reserve force for the Navy. That does not mean, however, that a Coast Guard variant of the LCS could carry and operate each of the Navy's combat mission packages; whether that was possible would depend on how much of the

space reserved for mission packages was used to make the changes required for the Coast Guard variant of the ship. (For example, the equipment needed for the countermine-warfare mission package is much greater than that required for the surface-warfare package, which relies heavily on the LCS's aviation and hull-mounted gun systems and therefore would not be substantially affected by using the LCS as a Coast Guard cutter.) But this option would allow the Coast Guard's medium-endurance cutters to operate in the same areas and under the same conditions as the Navy's littoral combat ships. Coast Guard cutters supporting Navy operations today would operate with more restrictions in the event of war because they were not built to a warship's standard of survivability.

A second advantage of this option is that it would eliminate the risk of starting a new ship program. After the cost increases incurred with the lead ships of the LCS and the NSC, the Coast Guard, under this option, would be entering a shipbuilding program that is about to start serial production; no new-design OPCs would be purchased. Historically, once the first few ships in a new program are produced, the costs of subsequent ships come down and become more predictable, and the program is less vulnerable to technical risk. Furthermore, if funding were available, this option could enable the Coast Guard to start building its OPC force sooner than 2015.

Disadvantages of Option 1. This option would have two distinct disadvantages. First, it would be somewhat more expensive to implement than the services' plans. Second, this option would make it harder for the Coast Guard to conduct its normal day-to-day activities of search and rescue as well as enforcement of immigration, fisheries, and environmental laws and regulations. The reduction in the desired range and endurance of an OPC that used the LCS monohull would be the most important consideration regarding those Coast Guard activities. Under its current plans, if the Coast Guard were to build an off-shore patrol cutter based on the design of the national security cutter, its overall deepwater force of eight NSCs and 25 OPCs could provide the same number of mission days per year at 500 nautical miles as Option 1. However, at 1,500 nautical miles, Option 1 would provide about one-third fewer mission days than the Coast Guard's plan—3,800 mission days versus 5,600 (see Figure 7).

This option presents other potential disadvantages that are more difficult to quantify. The LCS cannot operate as

Box 1.

Calculating the Total Life-Cycle Costs of Small Combatants Under the Services' Plans and CBO's Options

Although acquisition costs usually make up half or more of a ship's total life-cycle costs, it is still useful to consider the long-term effects that decisions about acquiring certain types of ship can have on other elements of the Navy's and the Coast Guard's budgets.[1] One platform can be more or less expensive to operate, or it may have a longer or shorter service life than a different choice. Additionally, such costs are spread over many years into the future. To compare the costs of platforms with different streams of expenses over many years, the future costs are discounted to their present value to account for the time value of money.[2]

To estimate the future life-cycle costs of different combinations of littoral combat ships (LCSs), national security cutters (NSCs), and offshore patrol cutters (OPCs), the Congressional Budget Office (CBO) separately projected the annual acquisition, personnel, and operating costs for each possible ship from 2009 to 2055. CBO assumed that the disposal cost for each ship would be zero, as ships of this type are typically given or sold to other countries when they are no longer needed by the United States. The separate costs for each ship under each option were then discounted to 2009 dollars for comparison.

CBO estimates acquisition costs using various historical analogies of prior Navy ship programs, adjusted for weight. The acquisition costs include research and design as well as construction expenses. Learning and rate effects, and real (inflation-adjusted) cost growth in ship acquisition are also part of CBO's cost analysis. The operating costs are based on CBO's models of fuel, maintenance, and personnel costs. These models use historical operating costs and "steaming" hours con-

tained in the Navy's VAMOSC system.[3] CBO projected fuel costs for the LCS, NSC, and OPC using historical data for selected Navy surface combatants adjusted for displacement. Average maintenance costs were determined by using an average of several types of Navy ships as well as by incorporating information from the Coast Guard on maintenance costs and fuel usage for Coast Guard cutters. However, while there may be variations in maintenance costs associated with differences in manning levels among the options, CBO has not captured those variations in the estimates. Personnel costs are based on the manning assumptions in Title X and CBO's model of the fully burdened cost of personnel.[4] Those personnel costs include pay, withholding taxes paid by the government, housing benefits, current and future health benefits, retirement benefits, tax advantages, and veterans' benefits. CBO updated each category using 2009 data and projected each category to grow at rates consistent with CBO's long-term economic projections.

For each option, the costs are estimated for each ship and each year of the analysis. The costs are then discounted to their present value in 2009 dollars using a discount rate of 3 percent. CBO used a discount rate of 3 percent based upon the long-term average of 30-year Treasury bonds. That rate was chosen over the Office of Management and Budget's official rate of 2.8 percent because it was thought to better represent the long-run time value of money based upon CBO's economic projections.[5]

---

1. 10 U.S.C. 2434. See Charles J. Hitch and Roland N. McKean, *The Economics of Defense in the Nuclear Age* (Cambridge, Mass.: RAND Corporation, March 1960), p. 138.

2. Office of Management and Budget, *Guidelines and Discount Rates for Benefit-Cost Analysis of Federal Programs*, Circular A-94 (October 29, 1992); Hitch and McKean, *The Economics of Defense in the Nuclear Age*, p. 207.

3. The Navy's Visibility and Management of Operating and Support Costs (VAMOSC) management information system collects and reports historical operating and support costs for the U.S. Navy's and the U.S. Marine Corps' weapon systems. See www.navyvamosc.com.

4. Congressional Budget Office, *Evaluating Military Compensation* (June 2007).

5. Office of Management and Budget, *Guidelines and Discount Rates for Benefit-Cost Analysis of Federal Programs*, Appendix C; Congressional Budget Office, *A Preliminary Analysis of the President's Budget and an Update of CBO's Budget and Economic Outlook* (March 2009).

Box 2.

Increasing the Antiship Missile Defenses of a Patrol Frigate Derived from the National Security Cutter

Constructing the national security cutter (NSC) to have the same degree of survivability as the Navy's existing Oliver Hazard Perry class frigates would entail a major redesign of the ship—which, in turn, would result in substantial additional costs. Nevertheless, adding two additional layers of antiship missile defenses to a naval version of the NSC, as is described in Option 2, would significantly improve the ship's ability to defend itself.

For approximately $260 million, the Navy could replace the Close-In Weapon System (CIWS) currently used on the national security cutter with the SeaRAM Mk-15 CIWS. Unlike the former system, which consists of a rapid-firing gun designed to engage subsonic antiship missiles at close ranges, the SeaRAM CIWS would incorporate a rolling airframe missile on the same physical space but provide the

ship with the ability to engage supersonic antiship cruise missiles out to 5 nautical miles. The SeaRAM system includes its own sensor suite—a K_u band radar and forward-looking infrared imaging system—to detect, track, and destroy incoming missiles.

An additional layer of antiship missile defense could be provided by installing the Mk-56 vertical launch system with Evolved Sea Sparrow Missiles (ESSMs) along with an Mk-9 Tracker/Illuminator system to detect, track, and engage antiship missiles. The ESSM can engage supersonic antiship missiles at a range of nearly 30 nautical miles. Installing 20 sets of a 12-cell launching system (which would carry 24 ESSMs), buying the missiles, and integrating the weapons with the ships would cost about $1.1 billion.

---

safely as the Coast Guard's cutters in sea state five. Thus, there could be rescue missions that the Coast Guard might not be able to perform in a sea state 5 environment if it had to rely on an offshore patrol cutter. In addition, Coast Guard cutters are designed to conduct two small-boat operations simultaneously (operations that are necessary for search and rescue missions as well as for drug and migrant interdiction). One limitation of the LCS is that it can launch and recover only one small boat at a time from its stern ramp.

There are no evident disadvantages to the Navy if the Coast Guard were to use the LCS hull for its medium-endurance cutter—other than the slightly slower procurement of the Navy LCSs.

Option 2: Support the Navy's Maritime Strategy by Reducing the Number of LCSs Purchased and Buying NSCs Instead

Under this option, the Navy would purchase only 30 LCSs—as opposed to the 55 called for in its existing plans—and buy 20 national security cutters to use as

patrol frigates. The Navy's version of the NSC would carry more extensive antiship missile defenses than the Coast Guard's version (see Box 2). In 2010, the Navy would decide which version of the LCS it intends to buy; nearly all of the 30 LCSs in its fleet would be the same version.[17] The service would begin purchasing the NSC-derived patrol frigates in 2011 and build them at a rate of two per year. The Coast Guard would continue with its plan to purchase 25 medium-endurance cutters as a scaled-down version of the high-endurance NSC.

The acquisition costs to the government of implementing this option would be less than under the services' plans. The combined costs of the Navy's LCSs and patrol frigates and the Coast Guard's cutters would be about $41.5 billion, about $6 billion less than the $47.1 billion projected for the services' plans (see Figure 6 on page 14).

---

17. There would be two exceptions: the lead and second ships of the type that are not selected. The lead LCS monohull was authorized in 2005, the lead trimaran in 2006, and the second ships of both types were authorized in 2009.

Figure 8.

Minimum Number of Refueling and Resupply Visits During a 180-Day Deployment, Selected Navy Ships

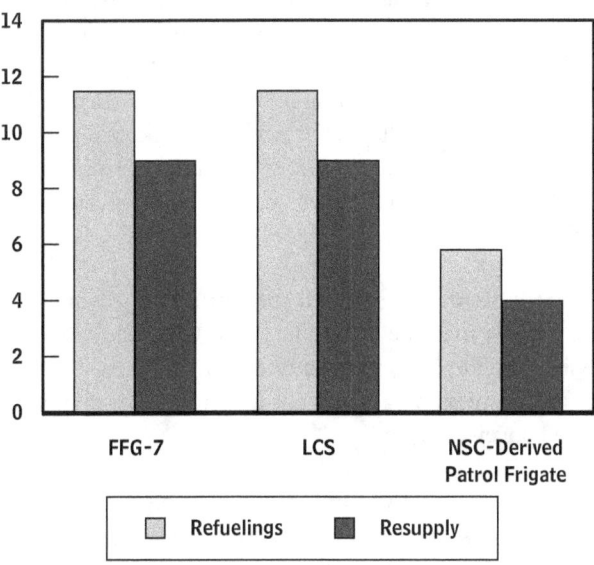

Source: Congressional Budget Office.

Notes: FFG-7 = Oliver Hazard Perry class guided-missile frigate; LCS = Freedom class littoral combat ship; NSC = Legend class national security cutter.

Ships can and will resupply more often, if convenient, depending on their operations.

The average price for the 78 small combatants under this plan would be about $530 million, compared with $570 million for 83 small combatants under the services' plans. Total life-cycle costs would be about 6 percent less under Option 2, totaling $92 billion in discounted dollars (in comparison with the $98 billion projected under the services' plans).

Advantages of Option 2. Implementing this option would bring several benefits to the Navy that might not be available with its LCS program. First, it would align the service's purchases of small combatants more closely with elements of its maritime strategy. The sea services' core statement of their missions and capabilities places about equal emphasis on peacetime operations: building naval partnerships with as many nations as possible around the globe; establishing global fleet stations; performing humanitarian operations; and conducting maritime security activities, such as counterpiracy and counterterrorism operations.[18] Such activities suggest the need for a low-

end-capability combatant that, much like the Coast Guard's cutters, has a long unrefueled mission range, high endurance, and is capable of many different types of independent operations. Those activities would not require the various mission packages and high-speed capability that are associated with the LCS program.

Second, with less than half the range and endurance of the national security cutter, the LCS requires more logistical support. That could be especially burdensome when it is operating in places far from where the Navy's logistics ships normally operate, such as off the coasts of southern Africa or South America. Using an NSC-derived patrol frigate would ease the strain on the Navy's own logistical support forces or reduce the frequency with which those ships would need to put into port to purchase fuel and supplies. Over the course of a 180-day deployment, the patrol frigate would require about half the number of refueling and resupply visits from Navy logistics ships or visits to regional port facilities (see Figure 8).

Third, the patrol frigate's ability to support a larger crew would make it easier for the Navy to perform maritime security and humanitarian operations. While the Navy generally is trying to reduce the number of people on its ships in order to save money, ships performing the type of operations envisioned here need to launch helicopters and accommodate boarding parties, often at the same time. With a core crew of only 40 and a total berthing capacity of 75, the LCS might not be as suited as the NSC-derived patrol frigate to conduct inspections and boardings that could become an increasingly routine part of the Navy's operations.

Disadvantages of Option 2. Nevertheless, this option would present certain disadvantages. Perhaps the most significant drawback is that even though the national security cutter comes equipped with some self-defense capability, it is not designed for use in combat operations. To compensate for that, the Navy could, for about $60 million per ship plus additional one-time costs, add more self-defense capability to the ships, such as the rolling-airframe missile and the Evolved Sea Sparrow Missile (which are antiship-missile missiles). CBO estimates that those improvements would cost about $1.4 billion. Even with those improvements, however, the patrol frigates would have higher acoustic and

18. *A Cooperative Strategy for 21st Century Seapower.*

Figure 9.

Number of Days Required to Deploy from San Diego to the Western Pacific, Selected Navy Ships

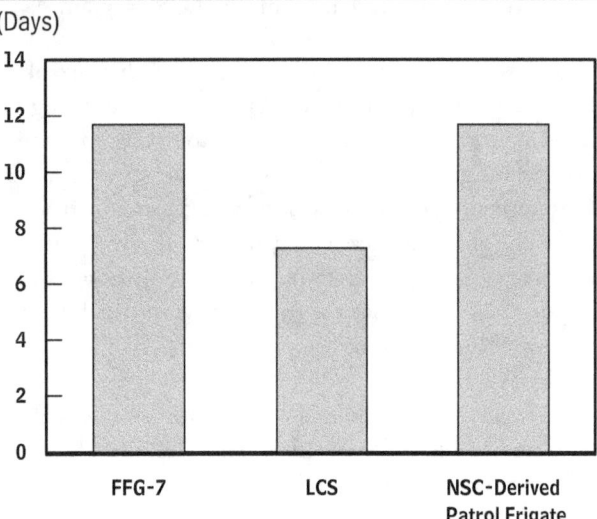

(Days)

Source: Congressional Budget Office.

Note: FFG-7 = Oliver Hazard Perry class guided-missile frigate; LCS = Freedom class littoral combat ship; NSC = Legend class national security cutter.

magnetic signatures than the LCS (thus making them more vulnerable to attack by diesel-electric submarines and mines); and the patrol frigate still would not be built to the same level of physical survivability as the LCS.

In addition, a patrol frigate cannot carry—at least not in their entirety—all the components of the mine-warfare and antisubmarine-warfare mission packages that the LCS will deploy. The NSC was not designed with enough internal space to accommodate those systems. (The NSC can carry all of the elements of the surface-warfare mission package.) Thus, if in a future conflict the Navy found that it needed more than the 30 LCSs in its fleet for combat operations, the 20 NSC-derived patrol frigates would not provide as much combat capability as would 25 LCSs. This option would not buy LCS mission packages for the NSC-derived patrol frigate. If the Navy decided that the patrol frigate should be prepared to carry elements of the mission packages, then the service would need to purchase those items, and the cost of Option 2 would be higher.

A third disadvantage is evident in combat scenarios where the need to quickly respond to a crisis or conflict is criti-

cal. If the Navy decided it needed to deploy more LCSs or patrol frigates from the United States to points overseas, the latter would take more time to deploy than the former. While the Navy has not yet revealed the LCS's maximum sustained speed, available evidence indicates that it is in the range of 40 to 45 knots. By contrast, the maximum speed of the national security cutter is 28 knots and that of a NSC-derived patrol frigate could be somewhat less depending on how the Navy decided to equip the ship. To illustrate the effect of speed, a patrol frigate deploying from San Diego to the western Pacific would require about 12 days, whereas an LCS could arrive in about seven (see Figure 9).

A fourth disadvantage is that the patrol frigate would be less flexible than the LCS. The latter, with its modular design and large internal volume, provides more flexibility to put mission packages on the ship that have not been conceived of today.

Option 3: Equip the Coast Guard's Deepwater Force with More National Security Cutters and Cancel the Offshore Patrol Cutter Program

This option would have the Coast Guard forgo designing and buying a medium-endurance offshore patrol cutter. Instead, the Coast Guard would buy 20 additional high-endurance NSCs, for a total of 28. The Coast Guard would begin purchasing additional NSCs sooner than is specified in its current plans, ramping up to two per year in 2011 and remaining at that rate of construction through 2022. The Coast Guard would purchase its last cutter in 2023, three years earlier than is called for under the service's plans.

The cost to the government to buy the ships under this option would be about $45.6 billion, compared with $47.1 billion under the services' plans (see Figure 6 on page 14). The average cost of the 78 small combatants under this option would be about $580 million per ship, compared with $570 million for 83 ships under the services' plans. The total life-cycle costs for the option would be $97 billion, compared with $98 billion under the services' plans.

Advantages of Option 3. A principal advantage of this option is that it would provide the Coast Guard's Deepwater force with slightly more mission days per year at points farther from the shore than is indicated in the service's plan. At 500 nautical miles from shore, this option would provide about 6,500 mission days per year

Figure 10.

Number of Days at Sea Provided Per Year by the Coast Guard's Current and Prospective Inventory of Cutters

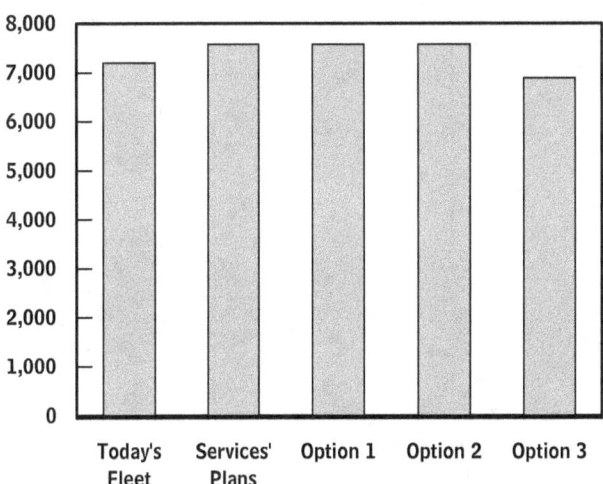

Source: Congressional Budget Office.

versus 6,900 under the Coast Guard's plan. However, at 1,500 nautical miles, this option would provide 5,700 mission days per year, compared with about 5,600 for the Coast Guard's planned Deepwater force (see Figure 7 on page 16).

By accelerating the NSC's construction and building more of them, this option would address the declining readiness of the Coast Guard's existing Hamilton class high-endurance cutters.[19] The Coast Guard would have 18 new cutters (including the first three NSCs) by 2020, compared with 11 under the service's current plan.

Another advantage of this option is that it would eliminate the risks associated with designing and building a new class of ship—specifically, the offshore patrol cutter. Both the Navy's and Coast Guard's small combatant programs experienced cost growth and schedule delays that significantly increased the amount of time and money needed to complete the first ships of those classes. While experience suggests that subsequent ships of those classes will not have the same problems—lead ships are notoriously difficult to bring to fruition—the Coast Guard

could see a similar result as it prepares to develop the still notional OPC.

Disadvantages of Option 3. A major disadvantage of this option is that it would provide fewer overall days at sea per year than would the Coast Guard's plans for its deepwater cutters. Specifically, the 28 national security cutters in this option's Deepwater force would provide a total of 6,900 days at sea based on the NSC's notional operating profile. The Coast Guard's plan for a mix of national security cutters and offshore patrol cutters—which, at 33 ships, would result in a larger overall Deepwater force—would provide about 7,600 days at sea, or about 10 percent more than under Option 3. By comparison, today's Deepwater force of high- and medium-endurance cutters provides about 7,200 days at sea per year (see Figure 10).

The effect of such a reduction in planned capabilities could make it difficult for the Coast Guard to fulfill all of its missions. For example, this option explicitly counters the recommendation made in a report issued by the RAND Corporation, which states that the Coast Guard will need twice the number of Deepwater assets that it currently plans to buy to meet both its traditional missions and its newly defined mission to protect the homeland (which emerged after the terrorist attacks of 2001).[20] That analysis was based on the number of mission hours the Coast Guard could be required to provide to perform its traditional and new missions.

A second disadvantage is that this option could compound the difficulties the Coast Guard would face in finding home ports for its national security cutters. In information provided to CBO, the Coast Guard states that many of the home ports that currently take the existing high- and medium-endurance cutters may not be able to accommodate the new national security cutter or, depending on the final design, the new offshore patrol cutter without significant, and potentially costly, modifications to the ports' facilities. Specifically, the NSC is about 10 percent longer and has a 10 percent greater draft than the Hamilton class high-endurance cutters.

Thus, the Coast Guard expects that it may have to expand those home ports that can accommodate the

19. Gordon, "Coast Guard to Assess Readiness of All High-Endurance Cutters."

20. John Birkler and others, *The U.S. Coast Guard's Deepwater Force Modernization Plan* (Santa Monica, Calif.: RAND Corporation, 2004).

NSC so that they can take more ships than they have previously, or the service may have to improve the port facilities in other locations. Those improvements could involve lengthening piers, dredging harbors and channels, and upgrading the electrical hookups at piers so that the cutters do not have to use their own generators when not at sea. Depending on the final design of the OPC, there could be a similar problem with the home-port facilities for medium-endurance cutters. The Coast Guard has not yet completed the analysis needed to determine where it will home port its new deepwater cutters and what upgrades, if any, will be required to all existing port facilities.

www.ingramcontent.com/pod-product-compliance
Lightning Source LLC
Chambersburg PA
CBHW081418170526
45166CB00010B/3388